Waste Plastics

Topical Conference at the 2021 AIChE Annual Meeting

Boston, Massachusetts, USA and Online
7 - 11 November and 15 - 19 November 2021

ISBN: 978-1-7138-5696-2

Printed from e-media with permission by:

Curran Associates, Inc.
57 Morehouse Lane
Red Hook, NY 12571

AIChE
120 Wall Street, FL 23
New York, NY 10005-4020

Phone: (800) 242-4363
Fax: (203) 775-5177

www.aiche.org

Additional copies of this publication are available from:

Curran Associates, Inc.
57 Morehouse Lane
Red Hook, NY 12571 USA
Phone: 845-758-0400
Fax: 845-758-2633
Email: curran@proceedings.com
Web: www.proceedings.com

TABLE OF CONTENTS

623077 Mechanistic Models of Polymer Degradation Chemistry

Abstract:

Pyrolysis is a promising method for resource recovery from plastic waste that thermally converts polymers in the absence of oxygen into valuable chemical feedstocks and monomer. It is also a detrimental process of degradation when polymeric materials are exposed to high temperatures. To provide further insight into polymer pyrolysis, a greater understanding of the mechanistic and kinetic details of the underlying reaction network is needed. To handle the complexity of mechanistic modeling of polymer degradation, we have developed both continuum and kinetic Monte Carlo (kMC) models. To facilitate creation of continuum models, we have formulated an automated modeling framework that assembles population balance models based on the method of moments from minimal user input. The population balances track the moments of a large number of radical and dead species, including monomer and small molecules that are relevant to degradation chemistry. We have applied this framework to study the degradation of polystyrene, polypropylene, polyisoprene, polystyrene peroxide and binary mixtures, and we are able to capture diverse experimental measures, including yields of individual low molecular weight products, as a function of reaction conditions. In order to solve the large models that are created, values of the rate coefficients for $O(10^5)$ reactions may be required. The approach that we have developed to specify rate coefficients is hierarchical, based on a combination of literature values, estimation methods, and computational chemistry.

626593 Tracking the Elementary Kinetics and Molecular Structures during Polyolefin Pyrolysis

Abstract:

Plastics have integrated into nearly all human activities since their invention 100 years ago, yet the recycling of plastics remains a difficult challenge. Herein, we examine the pyrolysis processes to chemically upcycle polyolefin plastics into monomers and other value-added chemicals. Polyolefin upcycling faces challenges that stem from the complex molecular composition of the feedstock, the large number of coupled chemical reactions and the dynamic changes that occur in the condensed phase chemical environment that forms, making it difficult to elucidate elementary kinetics and control the selectivity to products. These effects manifest in macroscopic phenomena such as differences in kinetics and changes in product distribution with changes in reaction temperature and feedstock composition. Understanding the molecular processes and developing a robust predictive kinetic model in polyolefin upcycling is crucial for the design and implementation of feasible plastic upcycling facilities.

Herein we present the development and application of an ab initio-based kinetic Monte Carlo and Molecular Dynamics (kMC+MD) simulation approach that can model the kinetics and product distribution of polyolefin pyrolysis. This approach tracks detailed atomic information including 3D coordinates of atoms and connectivity of chemical bonds in the feedstock molecules. It uses Stochastic Simulation Algorithm (SSA) to perform elementary reaction steps and uses classical MD simulations to follow the dynamics of the feedstock and the reaction environment as reactions proceed. The elementary step kinetics for the KMC simulations were established from density functional theory calculations.

The atomic-structural information is retained throughout the simulation, allowing us to follow molecular transformations and the local molecular compositions during the reaction. As such, the simulations capture the unique kinetic manifestations that would otherwise be lost in coarse-grained deterministic models. This KMC+MD simulation approach is used herein to examine the thermal pyrolysis reactions of polyethylene and polypropylene with some success in predicting kinetics and product distribution.

625020 Micropyrolysis of Polyethylene and Polypropylene: The Effect of Reactor Temperature and Vapor Residence Time on Product Distribution Prior to Bioconversion

Abstract:

The thermal degradation of plastics is a promising recycling technology that both addresses the waste management issues of landfill disposal and produces useful products. Primary thermal degradation of polymers usually yields large quantity of high molecular weight compounds with limited applicability. This makes secondary degradation necessary to improve the product's quality. We have found that the liquid product can be used as a feedstock for microbial biodegradation in enrichment cultures derived from environmental inocula. Oil-degrading microorganisms can colonize and break down the liquid alkenes, using the alkenes as carbon and energy source to produce microbial biomass or other value-added compounds.

In this study, pyrolysis vapors from polyethylene and polypropylene were subjected to secondary degradation using a new two-stage micropyrolysis reactor (TSMR) accessory to a commercial micropyrolysis unit (CDS Analytics). The reactor temperature (550-600 °C) and vapor residence time (VRT) (1.4-5.6 seconds) were varied. Pyrolysis products were analyzed using gas chromatography/mass spectrometry (GC/MS). The temperature and VRT variations showed a strong effect on the product distribution, comprising mostly alkene hydrocarbons with minor production of alkanes and alkadienes. At a low temperature (550 °C) and short VRT (1.4 s), a wide range of liquid (C5-C20 hydrocarbons) and wax products (C21-C30 hydrocarbons) are produced. Increasing temperature and VRT caused higher proportions of gas (C2-C4 hydrocarbons) and aromatics products are generated. Polypropylene was found to break down faster than polyethylene (HDPE and LDPE), producing lower molecular weight compounds under the same reactor conditions. A lumped kinetic model comprised of 10 reactions and 6 "species" was created to describe the plastic pyrolysis and to understand how the temperature and VRT effect product distribution into different product classes of compounds (plastic, wax, heavy oil, light oil, gas, and aromatics). Kinetic parameters, such as activation energy and frequency factor, are also reported.

623148 Lumped Product Kinetic Model of Plastics – Biomass Mixtures during Hydrothermal Liquefaction

Abstract:

Hydrothermal liquefaction (HTL) can be used to valorize waste plastics and biomass. Previous work indicates synergistic interactions between plastics and biomass and among plastics that increase the oil yields and influence product distributions. These interactions have the potential to reduce the process temperatures and improve the sustainability metrics for the process. Hence, understanding and modeling these interactions is pivotal to co-processing mixed (plastic – biomass) waste streams using HTL.

Herein to study these interactions, we take cellulose, polypropylene (PP), and polycarbonate (PC) as the model lignocellulose, polyolefin, and condensation polymer, respectively. We perform HTL experiments on each component alone (i.e., cellulose alone), each equi-mass binary combination (i.e., PP – cellulose (1:1 mixture)) and one equi-mass ternary combination (PP – PC – cellulose (1:1:1 mixture)) at 350 °C and six different batch holding times (1 – 60 min). We also vary the relative amount of each component in binary mixtures to analyze the ratio effects for each of the interactions through additional experimentation. A lumped product kinetic model is developed using these experimental results based on the pathway shown in the above figure.

The initial experimental analysis indicates synergistic interactions between PP – cellulose, and PC – cellulose that increase the oil yield. This means that mixing PP and cellulose and processing them together gives a higher oil yield than processing PP and cellulose separately. On the other hand, interactions among PP and PC are antagonistic and decrease oil yield. The increases in oil yield are accompanied by decreases in solid yield, while increases in solid yields accompany the decreases in oil yield. The degree of change caused by these interactions is further influenced by reaction times. Hence, kinetic models are ideal for understanding the interplay between reaction time and interactions and further understanding the reactions promoted/curbed by these interactions.

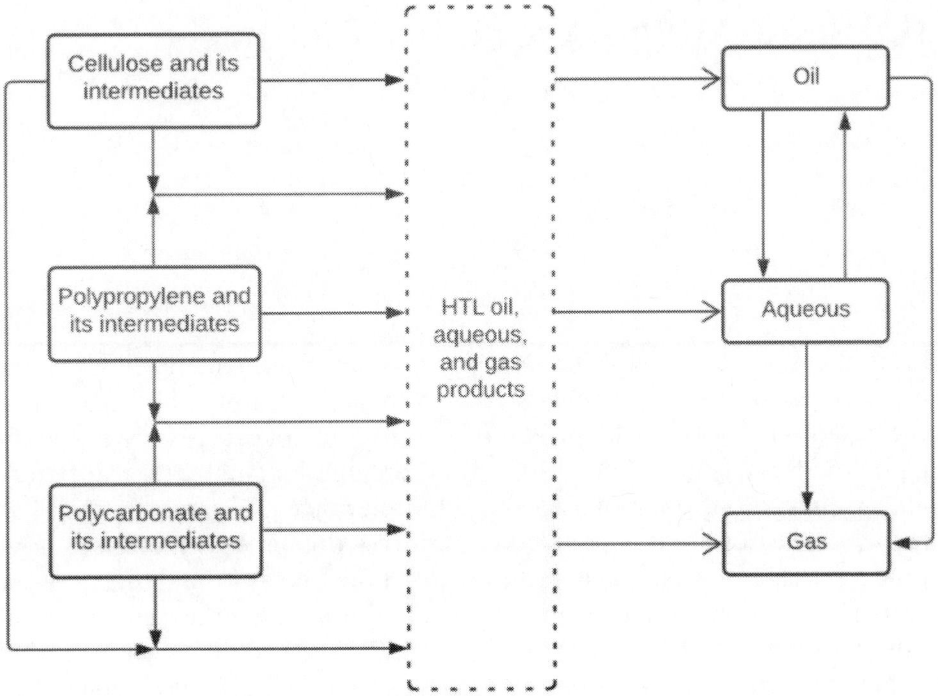

Fig: Reaction pathways for hydrothermal liquefaction of polypropylene (PP), polycarbonate (PC), and cellulose.

628288 Hydrothermal Process of the Ocean-Bound Plastic Waste into Value-Added Aromatic Chemicals

- Download/View Extended Abstract

Abstract:

Plastics are robust and versatile and thus have wide applications in several fields. However, the improper disposal of plastics, which are persistence in the environment, has made plastics waste a global issue. Particularly, recovering the plastic waste from the ocean is very costly. Conventional plastic recycling methods (*e.g.*, mechanical recycling) are challenged by their inferior efficiency, possible emissions of hazardous compounds, and little tolerance of contaminants. Currently, most mechanically recycled plastics are downcycled. To address this problem, in this study, an hydrothermal processes (HTP) was developed to convert the ocean-bound plastic waste (OPW), which are generally wet and has contaminants, into useful products under supercritical water. Based on the degradation temperature of the OPW feedstock (425°C), HTP tests were carried out at 425-475°C and 23MPa with reaction times of 0.5−6 h. It was found that above 78 wt% of OPW could be converted into oil products by the increased temperature (≥450°C) or elevated reaction times (≥2h). Different amounts of polypropylene mixed with OPW were found to reduce the severity of reaction parameters required for aromatics production. The fuel fractions of the oil phase within different boiling ranges were determined by TGA analysis. Over 60% of the oil can be used as naphtha (with the boiling point range of 25°C-200°C). The chemical composition of the oil products will be analyzed by Fourier-transform infrared spectroscopy (FT-IR) and gas chromatography–mass spectrometry (GC-MS). Integrated with elemental analysis of the oil products, a reaction mechanism will be proposed. Results from this study would contribute to the development of an efficient advanced recycling method for ocean plastic waste.

625667 Controlled Depolymerization and Upcycling of PVC

Abstract:

PVC is the 3rd most produced plastic worldwide. Rigid PVC is the common material of construction for most modern plumbing systems, and other forms of PVC find their way into a wide range of consumer products. The presence of chlorine atoms in PVC presents unique challenges for its disposal as the uncontrolled decomposition of PVC releases HCl and chlorinated organic compounds, presenting waste management and environmental problems. However, the design of chemical reactions and processes which control the decomposition of PVC affords unique opportunities for depolymerization and upcycling. This presentation will detail our efforts to date in the conversion of waste PVC into an array of value-added small molecules and polymer products.

629718 Selective Hydrogenolysis of Polyethylene and Polypropylene to Liquid Alkanes over Tunable Ruthenium-Based Heterogeneous Catalysts

Abstract:

The massive consumption of single-use plastics has led to a rapid accumulation of waste plastics in landfills and the environment, causing a global environmental and health crisis. Chemical upcycling has emerged as a method of producing higher value products from waste plastic feedstocks. The strong C-C bonds in common plastics such as polyolefins are particularly recalcitrant to degradation. Heterogeneous thermal catalysis enables targeted cleavage of bonds that can depolymerize polyolefins under milder conditions relative to other thermal deconstruction methods such as pyrolysis and gasification. We previously identified Ruthenium nanoparticles supported on carbon as an active and selective catalyst for the deconstruction of polyethylene via hydrogenolysis under relatively mild conditions (200-225°C, 20 bar H_2) to produce high yields of liquid n-alkanes. In this study, the differences in C-C bond activation of straight-chain (polyethylene) versus branched (polypropylene) hydrocarbons were investigated to understand the effect of branches on the mechanism and kinetics of C-C bond hydrogenolysis. Further, we hypothesized that nanoparticle-support interactions play a role in dictating hydrogenolysis selectivity, and that undesired methane formation could be suppressed by tuning the support acidity, nanoparticle size, and nanoparticle distribution. To this end, a series of Ruthenium-based catalysts on varying supports with different nanoparticle size and distribution were synthesized and characterized with TEM, XRD, BET, NH_3-TPD, H_2-TPD, and CO_2-TPD to identify the Ru dispersion, support structure, and acid-base properties. These catalysts were employed for the hydrogenolysis of polyethylene and polypropylene feedstocks to produce liquid n-alkanes and iso-alkanes, respectively. By tuning the support structure, acidity, confinement, and nanoparticle size, a liquid iso-alkane yield of 80% (C_8-C_{18} hydrocarbons) from polypropylene (M_w ~340,000 Da) was achieved. Further investigation into bifunctional Ru-based catalysts on acidic supports, confinement effects, and earth-abundant materials are currently underway, and may enable economical routes to produce high quality processible liquids from waste plastic feedstocks.

623610 Effect of Chain Branching on the Hydrogenolysis of Heptane Isomers over Metal Catalysts: Insights to Hydrogenolysis of Polyethylene.

Abstract:

The ever-increasing rate of production of plastics, most especially single-use plastics, have plagued our world with significant amounts of pollutants due to the limitations in plastics recycle. This has spurred the interests of researchers to design catalysts that selectively upcycle plastic wastes into value-added materials. One of the major techniques being studied is the catalytic hydrogenolysis. Due to the similarities in the chemical structure and relative inertness of alkanes and polyethylene, which accounts for most plastics produced, it has been suggested that the same mechanism may guide their hydrogenolysis.

In this work, we investigated and compared the mechanism of the hydrogenolysis of linear and several branched isomers of heptane over Pt(111) and other Pt surfaces to smaller hydrocarbons and intermediates, using density functional theory (DFT) calculations. As claimed by previous work on alkane hydrogenolysis, we found that C-C bond activations have to be preceded by a series of dehydrogenation steps, which reduces their degree of saturation and allow for Carbon-Metal bonding. In the absence of dehydrogenation steps, C-C bond cleavage has a high energy barrier.

Next, we performed a mean-field microkinetic model for the hydrogenolysis of heptane isomers to establish reaction pathways, determine the rate for each elementary step, and determine the rate controlling step. Finally, we are in the process of developing models for extrapolating information from alkanes to the hydrogenolysis of polyethylene chains.

621498 Chemical Upcycling of Polyethylene Using Heterogeneous Alkane Cross-Metathesis

Abstract:

High/low density polyethylene (PE) represents the largest portion of the plastic waste stream and a major challenge in chemical recycling due to stability of the C-C bonds that make up the backbone of these polymers. In the effort of developing effective and economical chemical upcycling processes, catalytic alkane metathesis chemistry comprising tandem (de)hydrogenation–olefin metathesis has been investigated as a novel approach due to two favorable factors for economic viability of the recycling process: its moderate operating temperature and absence of a reactive gas—*i.e.*, H_2. However, long reaction times, poor catalyst stability, a narrow operating temperature range, and the need for expensive metal components (Re) remain important challenges. Here, we demonstrate that WO_x/SiO_2 (WS) catalyst can successfully substitute rhenium oxide for the olefin metathesis stage of the tandem reaction in n-hexadecane (n-C_{16}; a surrogate of PE) alkane-metathesis reaction at 300 °C and shorter reaction time (2 hours) in a 50mL batch reactor.

The high olefin metathesis reactivity of WS catalyst is revealed by the results of the 1-hexadecene (1-C_{16}=) metathesis reaction. The outcome is a broad distribution of olefin products in which selectivity toward olefins of carbon number lower and greater than 16 is 45% and 55%, respectively (Figure 1a). For the alkane metathesis reaction, the simultaneous use of WS and Pt/γ-Al_2O_3 (PA) proved effective for converting a mixture of n-C_{16} and n-C_7 to alkane compounds of lower carbon number (< C_{16}) with a selectivity of 85% (Figure 1b). A product yield of 165% (n-C_{16} basis) establishes that n-C_7 participates in the reaction as feedstock through self or cross-metathesis and, consequently, decreases the average molecular weight of overall alkane products. These results validate the premise that the alkane metathesis reaction using tungsten oxide is an attractive approach to PE upcycling.

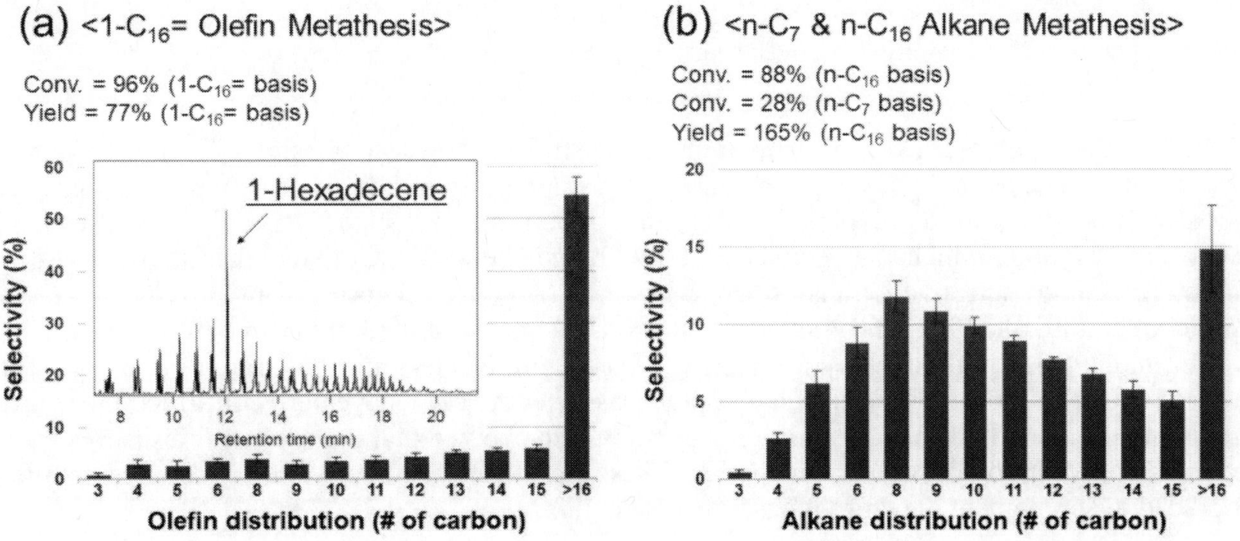

Figure 1. (a) Product distribution of olefin compounds after olefin metathesis reaction of 1.56 g of 1-hexadecene feedstock ($1\text{-}C_{16}^{=}$) with n-heptane ($n\text{-}C_7$); $n\text{-}C_7/1\text{-}C_{16}^{=}$ = 9.8 (mol/mol). 800 mg WO_x/SiO_2 (2 wt %) was used as catalyst and 1 g of 4A adsorbent was added. The bars represent as an average of reaction triplicate with error bars shown as standard deviation. Reaction condition: 300 °C, 30 bar N_2, 2 hours. (b) Distribution of alkane products after alkane metathesis reaction of 1.58 g of n-hexadecane feedstock ($n\text{-}C_{16}$) with n-heptane ($n\text{-}C_7$); nC_7/nC_{16} = 9.8 (mol/mol). 500 mg $Pt/\gamma Al_2O_3$ (1 wt %) and 800 mg WO_x/SiO_2 (2 wt %) were used as catalysts and 1 g of 4A adsorbent was added. The bars represent as an average of reaction quadruplicate with error bars shown as standard deviation.
Reaction condition: 300 °C, 30 bar N_2, 2 hours.

629762 Selective Production of Gasoline Range Iso-Alkanes from Polyolefin Plastics

Abstract:

Plastics made of polyolefins are widely used in daily life, such as low-density polyethylene, high-density polyethylene, and polypropylene. Despite a small portion of high-density polyethylene is recycled, most of the used polyolefin plastics are not currently recycled. These plastic wastes are great hydrocarbon sources which can be cracked to fuel range chemicals. Our study has shown that these polyolefins can be selectively transformed to gasoline range iso-alkanes (mainly C7, C8 and C9) at a mild temperature of 350 °C and 1 standard atmospheric pressure, with the help of Y zeolites, which are commonly used in fluid catalytic cracking of crude oil. Product analysis from Fourier-transform infrared spectroscopy and gas chromatography-mass spectroscopy have both confirmed the selective production of the iso-alkanes. In addition, the study also showed that the used zeolites can be regenerated by air combustion and reapplied. These iso-alkanes could be directly adopted or further upgraded to supplement the current gasoline production.

625858 Bridging the Gap between Model Alkanes and Polyethylene Waste Hydrocracking

Abstract:

Plastics have become ubiquitous in modern society. Unfortunately, the current management of plastic waste (PW), based mainly on landfill disposal, is unsustainable. Chemical recycling of PW, particularly of polyolefins (PO), to liquid fuels and lubricants, is promising for generating value while diverting it from landfills.

Hydrocracking is widely exploited in refineries to valorize waxes and heavy oils (C_{20}-C_{50+}) and is attractive for the chemical recycling of PO. Hydrocracking catalysis is well understood for small alkanes (C_6-C_{12}). However, it is unclear if these model compounds' mechanistic concepts can be directly applied to PO systems.

Platinum on select oxides has been demonstrated to be active and selective hydrocracking catalyst. Moreover, its physicochemical properties are well established, making it an ideal candidate for fundamental studies.

The current work establishes a mechanistic framework for PO hydrocracking and identifies reactivity descriptors to close the gap between model alkanes and PO. We assessed the effects of each catalyst component loading on low-density polyethylene (LDPE) as a function of reaction conditions. Our results show that the catalyst composition and structure play a significant role in LDPE hydrocracking. We demonstrate that we can control the distribution to larger alkanes and enhance the branching in the residual polymer. Moreover, we identify key similarities and differences between the hydrocracking behavior of small alkanes and LDPE and propose a new macromolecular hydrocracking mechanism that departs from small alkanes. We perform extensive characterization of the catalyst, the products, and the residual polymer to provide structure-reactivity relations.

624859 Application of Automatic Fragment Modeling in Plastic Pyrolysis

Abstract:

Detailed kinetic models are well developed for predicting light hydrocarbon systems. However, for more complicated modeling processes such as waste plastic pyrolysis, new approaches are required. To overcome difficulties including exponentially growth of number of possible isomers and reactions or the uncertain ability of extrapolation for lumping methods, we develop a fragment-based framework called Auto-Fragment Modeling which targets to build detailed kinetic models for heavy hydrocarbons. This proposed fragment-based method treats large molecules as independent fragments. Each fragment represents different part of the original molecules which share common structures. It is assumed that similar structures will conduct similar reactions.

This method combines with open-source software Reaction Mechanism Generator (RMG) software package and estimates rate coefficient and thermochemistry parameters derived from high-accuracy quantum chemistry. Previously we demonstrated several examples to show promising results by capturing important chemistry as well as product distributions. The new application system elucidates the reaction mechanisms leading to several of the functional groups formed in this important process. Some technical challenges that still need to be overcome to allow accurate predictions for systems containing macromolecules are identified.

625352 Catalytic Depolymerization of Waste Polyolefins By Induction Heating: Selective Alkane/Alkene Production

Abstract:

The global production of plastics, 400 mmt in 2015, is rising at >4%/yr. Nevertheless, the recycling rate of waste plastics is unsustainably low, <10% in the U.S. The current approaches to recycling plastics have many limitations. For example, plastics pyrolysis is limited by an unwieldly product distribution and high operating temperatures. In this work, we depolymerized polyethylene (LDPE, HDPE) to useful products by utilizing efficient energy input from RF induction heating, coupled with a heterogenous catalyst.

We have selectively depolymerized, without added H_2, low- and high-density polyethylene to C2-C20+ alkanes/alkenes by utilizing efficient energy (lower fluid temperature) RF induction heating, coupled with dual-functional heterogenous catalysts consisting of Fe_3O_4 nanoparticles and Ni-based catalysts. The Fe_3O_4 was used to locally generate heat when exposed to the RF field (<65 mT). Initial results indicate that nickel-exchanged zeolite catalysts are more selective to light olefins with fewer aromatics and minimal coke formation, while nickel supported on CeO_2-ZrO_2 or CeO_2 is more selective to C7-C18 alkanes/alkenes, but with some coke formation. LDPE conversions up to 95% was obtained with Ni supported on Fe_3O_4 at 600 A, 2 h, with minimal aromatics, coke and light gas formation. The difference in selectivities is associated with two possible depolymerization mechanisms, a reverse Cossee-Arlman mechanism or a random cleavage process. The depolymerization process was also tested on commercial LDPE (grocery bags) using the Ni on Fe_3O_4 catalyst, resulting in similar conversion to liquids and gases (~48%) as for virgin LDPE but reduced coking under the same reaction conditions.

627069 Low-Temperature, Energy-Efficient Hydrocracking of Polyolefins to Fuels

Abstract:

Plastics waste (PW) contamination of the environment is a major global ecological and economic challenge. Polyolefines, including polyethylene (PE) and polypropylene (PP), which constitute about 57% of all PW are hard to recycle using currently offered technologies due to the rigidity of C-C bond in the polymer backbone. A promising approach to break PE and PP down to smaller fuel range hydrocarbons in an energy-efficient way is hydrocracking [1]. This study aims to develop an optimal hydrocracking catalyst. Another goal is to provide guidelines in tuning process selectivity from gasoline to lubricants by changing catalyst properties/composition or by altering reaction conditions.

Hydrocracking was performed in a batch reactor at 10-50 bar H_2 pressure for 1-12 h using different virgin and waste polymers as a feedstock. Platinum (Pt) on tungstated zirconia (WO_3/ZrO_2) was found to be active in the conversion of PE with a molecular weight of 250 kDa to C_7-C_{12} alkanes with a fraction of branched isomers close to 90%. When mixed with HY zeolite activity increases 4-fold with a shift in selectivity to lighter C_5-C_7 hydrocarbons. Detailed investigation of reaction network showed that zeolite plays an important role in cracking of primary olefins originated from $Pt/WO_3/ZrO_2$. The acid site density and mesoporosity of the zeolite also significantly affect performance. Lower acidity together with enhanced micropores accessibility favor less deep cracking. The catalysts showed nearly complete reusability after calcination. Fuel-grade hydrocarbons yield reaches 80-85% at 250-275 °C within 2-4 h of reaction. Major mechanistic insights allow us to propose further development of plastic waste hydrocracking.

627663 Understanding Chemical Recycling of Step-Growth Polymers Using Kinetic Monte Carlo Approach

Abstract:

Step-growth polymers like polyethylene terephthalate (PET) find their applications in the production of various materials. As post-consumer plastic waste poses a serious environmental threat, efforts to design novel bio-based materials and efficient recycling strategies are underway. Towards this end, it is crucial to understand the depolymerization chemistry to optimize the recycling process. The existing kinetic studies provide little information on the influence of properties like structure, degree of polymerization, or molecular weight on the monomer recovery.

The current study aims at developing a detailed mechanistic model to unravel depolymerization pathways of PET using the kinetic Monte Carlo (kMC) approach. The model comprises three steps, viz. polymer reconstruction, reaction network generation, and obtaining product distributions using the kMC framework. For simple polymers like PET, the polymer chains are comprised of several ester bond linkages. Automatic network generator creates the list of possible reaction events based on a set of defined reaction families that characterize the types of bond cleavage events and any possible side reactions, which allow the formation of intermediate and LMWP species. Finally, the kMC framework simulates these reactions as single events occurring at discrete time steps.

First, Schulz-Flory distribution method was used to determine the maximum chain length and the corresponding number of PET chains in the feed. Further, studies in the absence of a catalyst were used to provide a baseline, fitting rate coefficients for the key bond cleavage events. The neat chemistry was then combined with catalyzed pathways to evaluate the catalyst's effect on the molecular weight decay and the temporal yield of LMWP. Molecular weight information as a function of time in a batch reactor was modeled, and the yield of low molecular weight products was monitored. Finally, the mechanistic model was validated using the experimental data available from the literature on glycolysis of PET.

626153 Targeting Fuel-Ranged Aromatics through Co-Pyrolysis of Polystyrene and Polyethylene with H-ZSM5

Abstract:

The objective of this work is to upcycle low-density polyethylene (LDPE, resin code 4) and polystyrene (PS, resin code 6), which are currently unrecycled, for producing gasoline ranged aromatics. Common plastics for LDPE and PS are grocery bags and packaging foam, respectively. Typically, both plastics are either combusted along with other municipal waste inside power plant or transported to be landfilled. Due to environmental concerns resulted from current treatments, chemical upcycling could be promising since it could achieve almost total conversion of both plastics to fuel. In this study, fluid catalytic cracking catalyst H-ZSM5 were adopted for the catalytic co-pyrolysis of LDPE and PS. Hydrogen-rich LDPE could supplement the pyrolysis of PS to increase the yield of alkyl-benzene. In addition, the products from LDPE could diversify the aromatics from PS pyrolysis.

Experiments were carried out with the three independent variables: the presence of catalyst, catalyst acidity and catalyst ratio. H-ZSM5 used in this work varied in three silica to alumina ratios, 30, 80 and 280, corresponding to the acidities of each catalyst. PS and PE were tested individually and as a mixture (1:1 mass ratio) with and without the presence of H-ZSM5. Simulated cracking using thermogravimetric analysis (TGA) showed that the catalytic cracking had reduced the required temperature from 490 °C of non-catalytic cracking to 340 °C, for almost total conversion. Online Fourier-transform infrared spectroscopy (FTIR) of the flue gas from TG furnace showed that production of aromatics increased with the addition of PS in PE. For the products PS and mixed plastic samples, in the FTIR spectra, a peak at 700 cm^{-1}, attributing to benzene and derivatives, was found. This particular peak was absent in the PE products. The mass spectroscopy analysis of fixed bed reactor showed a more diversified aromatics produced from co-upcycling of PE and PS.

627538 Synergetic Effects of Co-Pyrolysis of Hdpe and Various Biomass Feedstocks By ZSM-5 and MCM-41/ZSM-5 Composites

Abstract:

The production of renewable fuels has become necessary as fossil fuels continue to deplete past the point of no return. Catalytic fast pyrolysis (CFP) of biomass to high-value liquid fuels has emerged as a promising solution. However, CFP is hindered by the low hydrogen to carbon (< 1 H:C) effective ratio of biomass, which results in high coke and low oil yields. Co-pyrolysis of biomass and hydrogen-rich materials, such as plastic, can increase the overall H:C ratio, thereby increasing oil yields. Over 4.8 million tons of plastics have entered the ocean since 2010, making co-pyrolysis of waste plastic and biomass an effective waste utilization method while increasing pyrolysis effectiveness.[1]

In this work, catalytic fast co-pyrolysis using a Py-GC/MS system was performed on real feedstocks (miscanthus × giganteus and food waste) and model compounds (cellulose, xylan, lignin and protein) mixed with high density polyethylene (HDPE) to elucidate the role of feedstock composition on oil yield in co-pyrolysis with ZSM-5 and composite MCM-41/ZSM-5 catalysts.

Results presented in figure 1 show the synergetic effect of HDPE in co-pyrolysis with miscanthus. The overall liquid and major liquid product class yields are increased at all mixing ratios. Notably, the phenol content of co-pyrolysis oils is lower than that of pure biomass corresponding to a lower oxygen content. The optimum mixing ratio of HDPE to miscanthus was found to be 1:3 corresponding to the largest increase in total liquid carbon yield of 12%. This work aims to show the influence of biomass molecular makeup on performance in co-pyrolysis with HDPE in order to effectively tailor feedstock composition for greater yields.

References

(1) Jambeck, J. R.; Geyer, R.; Wilcox, C.; Siegler, T. R.; Perryman, M.; Andrady, A.; Narayan, R.; Law, K. L. Plastic Waste Inputs from Land into the Ocean. *Science (80-.)*. **2015**, *347* (6223). https://doi.org/10.1126/science.1260352.

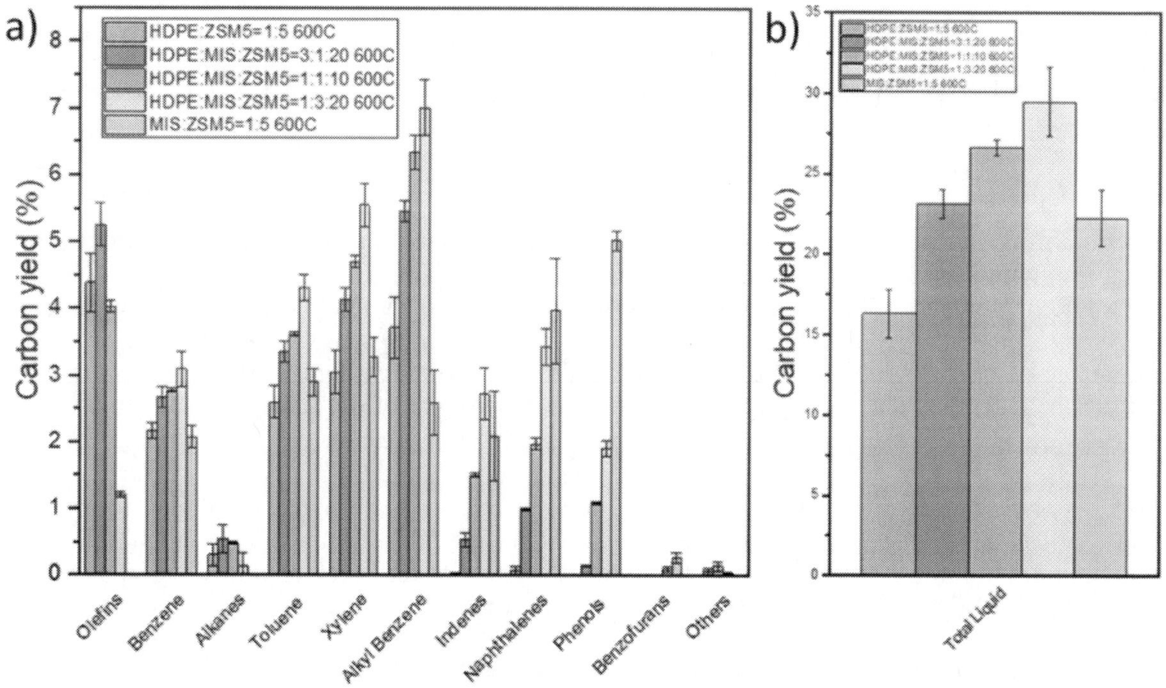

Figure 1. Carbon yields from co-pyrolysis of HDPE and miscanthus at various mixing ratios for a) the major liquid product classes, and b) the total liquid products

621762 Eastman in the Circular Economy – How Life Cycle Assessment (LCA) Enables Commercial Scale Molecular Recycling

Abstract:

The global plastic waste crisis and climate change are two of the greatest challenges of our time. The world desperately needs a materials revolution to help address both. Eastman continuously strives to enhance the sustainability of our products and manufacturing processes, and uses life cycle assessment (LCA) as a holistic approach to quantify the potential environmental impacts of a product or activity throughout its life cycle — from raw material and resource extraction, to manufacture, consumer use, and end of life. Results from LCA studies can be used to inform decisions at many levels, including material selection, design considerations, corporate strategy, and policy. Eastman is advancing circular economy solutions for plastics through innovative molecular recycling technologies that transform difficult-to-recycle plastic wastes into raw materials for producing new specialty plastics and fibers with no compromise in quality or performance. The plastic waste streams include pigmented materials, mixed plastics, post-consumer carpet, used clothing, and reject streams from traditional mechanical recycling.

This talk will focus on Eastman in the Circular Economy: how molecular recycling technologies are enabling Eastman to lead in addressing plastic waste issues and how LCA has enabled decisions that address the plastic waste issue while simultaneously improving overall carbon footprint. The talk will discuss at a high level the life cycle environmental impacts of two commercial-scale Eastman facilities for advanced molecular recycling of waste plastics in the US – Polyester Renewal Technology (PRT) and Carbon Renewal Technology (CRT). PRT is a depolymerization process for difficult-to-recycle polyester wastes; CRT is a reforming process for recycling of mixed plastic wastes. Both technologies break plastic waste into molecular building blocks which are directly used to produce a variety of specialty plastics and fiber materials. Eastman is dedicated to sharing our learnings with other value chain partners to help our world meet important greenhouse gas reduction goals while building and supporting a more circular economy.

634190 Creating a Pure Planet by Transforming Recycling

Abstract:

PureCycle Technologies strives to create a Pure Planet by commercializing the solvent-based polypropylene extraction technology invented by Procter & Gamble. Polypropylene (PP) is used in applications ranging from automobiles to food packaging and durable goods to textiles. Each PP application requires a unique formulation of solids, elastomers, colors, and additives to achieve the desired properties for the final product. This application diversity makes recycling polypropylene at scale elusive to traditional recycling methodologies due to the challenges of blending the different components of the recycled products. Polypropylene is the most produced polyolefin in the world at approximately 150 billion pounds per year and is growing at near double GDP rates, yet remains among the lowest recycling rates in the world at less than 1%. This presentation will frame the global plastic waste crisis, how new thinking and novel technology is required to solve this global problem, and how PCT intends to bend the status quo to change the narrative on recycling.

629230 Computational Prediction of Polymer Solubilities and Its Application in Multilayer Plastic Film Recycling

Abstract:

Billions of pounds of multilayer plastic films are produced annually as packaging materials. These films are made of distinct polymer layers, which makes it a challenging task to recycle them. Recently, we developed a strategy called solvent-targeted recovery and precipitation (STRAP) that can deconstruct and recycle such films through a series of solvent washes. It utilizes the solubility difference of the polymers to selectively dissolve them in carefully chosen solvent systems. One key step in designing a STRAP process for a target film is to identify selective solvent systems and operating temperatures, which could be prohibitively time-consuming using experimental screening alone. Here, we report a computational scheme for polymer solubility prediction and solvent selection that employs Hansen solubility parameters, molecular dynamics simulations, and the COnductor-like Screening Model for Realistic Solvents. These tools take molecular interactions and conformational information into consideration to yield quantitative, temperature-dependent solubility predictions of polymers in pure solvents and solvent mixtures. We demonstrate that this computational scheme can efficiently screen appropriate solvents to guide the design of solvent-mediated processes for multilayer plastic film recycling.

631196 Calibration of a CFD-DEM Fluidised Bed Simulation for Plastic Recycling Using Positron Emission Particle Tracking

Abstract:

Industrial-scale plastic recycling using pyrolysis in fluidised beds is an emerging technology aiming to address the ever growing plastic waste crisis. Whilst we developed a basic understanding of fluidised beds with a single solid phase, such as the fluid catalytic cracking in modern oil-refineries which have been used for decades, binary systems which introduce a non-identical second solid species often lead to difficulties in reactor design. Uneven particle distributions, segregation and hot-spots lead to not only product issues but also safety issues which need to be understood and removed.

The Discrete Element Method (DEM) does, when calibrated properly, reproduce the behaviour of moving granular material very well. Coupled with the well developed computational fluid dynamics (CFD) it provides a powerful tool in predicting the behaviour of a fluidised granular system such as fluidised beds.
These systems however need to be calibrated and even though the conventional techniques of calibration for DEM models such as shear cells or angle of repose measurements can lead to quantitative granular properties, the characterisation and calibration of the fluid phase is currently much less precise.

We use positron emission particle tracking, a Lagrangian particle tracking technique, to directly compare coupled DEM-CFD simulations of fluidised beds with real world fluidised beds. The rich, high-resolution data yielded by PEPT allows us to fine tune drag models and particle parameters in a significantly more precise manner than is possible using conventional measurements such as pressure drop alone.

631881 Use of Waste Plastic in Asphalt Binder: Theoretically and Experimentally Justified Identification of Compatible Blends

Abstract:

Previous attempts to add waste plastic to asphalts were not met with great success primarily due to the incompatibility of common plastic and asphalt cement, which resulted in poor performance. If the compatibility between waste plastic and asphalt binder can be systematically understood and improved, then large amounts (millions of tons annually) of previously land-filled waste plastic can be repurposed in asphalt binder. Here, we fundamentally investigate the compatibility of waste plastic and asphalt using a combination of atomistic simulation, coarse-grained simulation, and experimental validation to provide guidelines for predicting the compatibility between asphalt and waste plastic. Waste plastics such as HDPE, LDPE, PVC, PET, and PS are considered, along with some approaches to improve compatibilitzation such as crosslinking with ground tire rubber, aminolysis of PET, and wax residue addition. These materials and approaches are considered in terms of fundamental polymer science concepts.

633557 Thiol-functionalized Hyper-Cross-Linked Milk Protein polymers for Mercury Removal

Abstract:

Hyper-cross-linked organic polymers (HCPs) have recently attracted a lot of attention as their large surface area and strong interactions with other molecules making them highly appropriate for different applications. The production of this type of microporous polymers is usually accomplished by employing pricey reagents. As an inexpensive way of synthesizing HCPs scaffold, we create one step Friedel–Crafts alkylation reaction with milk caseins, existing in high quantity and exhibiting unique polymer properties. To optimize the surface area of the polymer, the ratio of the catalyst (AlCl3) to the cross-linker (2,4,6-Trichloro-1,3,5-triazine) was studied. Then, the polymer was decorated with sulfur functional groups to become a suitable platform for mercury separation from the water and gas phases. Our approach together with characterization techniques (FTIR, ICP, XPS, BET) validated that the integration of thiol-groups into HCP leads to remarkable de-mercuration performance. Thiol-functionalized casein polymer scaffold demonstrates excellent capacity and selectivity for trapping heavy metal ions. The high capabilities of developed sustainable adsorbents from waste streams make them comparable to current state-of-the-art nanoporous materials.

623747 Effect of Selective Impurities on Carbon Capture from Biogas Using Deep Eutectic Solvents

Abstract:

Anaerobic digestion (AD) of food wastes is one of the most promising pathways to convert wastes to value added fuels (e.g., biogas) and products (e.g., fertilizer). As food wastes are heterogeneous and their composition changes from household-to-household, the resulting biogas composition varies significantly. Biogas from food wastes often contain 40-60 % methane (CH_4), 40-60 % carbon dioxide (CO_2), and 1-5 % impurities. Although the actual concentration may vary from the source of biomass and AD operating conditions, most common impurities found in the biogas are hydrogen sulfide (H_2S), ammonia (NH_3), octamethyltrisiloxane, and decamethylpentasiloxane. carbon capture and removal of the impurities are required prior to use or injection into a natural gas pipeline. CO_2 from the gaseous stream can be separated by physical (ionic liquid) or chemical solvents (amine-based solvents). Amine-based chemical solvents have higher adsorption capacity but are often expensive to regenerate and toxic for the environment. Physisorption solvents are generally expensive and their adsorption capacity is lower, meaning that significantly higher amounts of physisorption solvents are required. In the context of carbon capture, the impurities are likely to play a significant role in adsorptive capacity of the solvents due to an influence on the molecular environment.

In this study, deep eutectic solvents (DES) are evaluated to upgrade biogas. The benefit of DES is that they are made from the physical mixing of a hydrogen bond donor (HBD) and a hydrogen bond acceptor (HBA), generally made from a quaternary salt and acid/alcohols. In this study, a conductor like screening model (COSMO) is adopted to screen suitable HBA's and HBD's for biogas upgrading based on hydrogen bond properties using a triple-zeta valence polarization (TZVP) basis set. This model is evaluated compared to literature experimental studies to verify integrity of the results. Of the five DES studied, it was found that quaternary-octyl ammonium bromide with decanoic acid (N_{8888}Br: DA) showed the highest CO_2 adsorptive capability. However, the choline-chloride (ChCl) based solvents had a much higher selectivity for CO_2 over CH_4. Through an activity coefficient analysis, it was determined that choline chloride-based DES are much more affected by the presence of contaminants than the ammonium salts with regards to adsorptive capabilities. It has been observed that the HBD is the main determinant of contaminant sensitivity rather than alkyl chain length. Due to these several trade-off factors apparent from the results, it is determined that the concentration of impurities should be accounted for when determining an appropriate DES for carbon capture from biogas.

629073 Design, Simulation, and Validation of 3-D Printed Hydrocyclones for Microbial Cell Concentration

Abstract:

Concentration of microbial cells is a key step in bioprocessing and laboratory operations. At the laboratory scale, concentration is typically conducted in centrifuges, however, this requires energy and consumables, and is not compatible with autonomous, continuous operation. Hydrocyclones are promising alternatives because the only energy input is to operate a pump, there are no consumables, the hydrocyclone requires no labor after setup, and it can operate continuously. This study provides a parametric open source design of a 3-D printable hydrocyclone with interchangeable parts for ease of testing different hydrocyclone dimensions during experimental optimization. The hydrocyclone was simulated using COMSOL Multiphysics® software using the k-ε turbulence physics to simulate rotational fluid flow and coupled with the particle tracing module to track individual cells or flocs. The simulation was able to replicate the rotational vortex and central air core and could track individual cells to the hydrocyclone overflow and underflow. The simulation was used to estimate the concentration efficiency for different hydrocyclone geometries. Simulations were experimentally validated for *E. coli* cultures mixed with varying concentrations of food-grade flocculant to increase the size of particles in solution and improve cell concentration efficiency.

629770 Switching Bioprocesses into Zero-Waste Virtual Power Plants By Flexible Operation of Electrochemical pH-Swing Extraction Processes

Abstract:

Carboxylic acids are important platform molecules that enable a sustainable chemistry based on renewable resources. [1] Although the biochemistry of the production is well established the successful industrial application is still challenging. As an example the US-Canadian company Bio-Amber, once considered a rising star for their ambition to establish a fermentation based production route for succinic acid had to file for bankruptcy. [2] This demonstrates that up to today advantages in environmental metrics cannot sufficiently compensate an economic disadvantage of the bio-based production. In order to make bio-based chemistry successful a holistic process development is required and must strive to achieve high yields in the fermentation steps, a low-cost and low-waste separation process and exploit all opportunities to harvest every added value feasible in the process.

We developed an electrochemically driven pH-swing extraction process for the recovery of succinic acid from aqueous solutions. [3 4] This concept overcomes the issue of salt waste production otherwise encountered during the separation of bio-based succinic acid. [5] Instead of waste generating acid and base this technology uses electricity to drive the pH-swing extraction. The calculated electricity costs required to operate the electrochemical pH-swing extraction with the current prototype equipment are comparable to the costs for acid, base and waste disposal in state of the art process designs. A more detailed techno-economic assessment of the operational expenses (OPEX) of our electrochemical separation concept and four other process designs proposed in the literature was conducted with ASPENplus. The study revealed a cost advantage of the electrochemical processing route compared to other downstream options. Beyond the promising OPEX values determined for the electrochemical pH-swing process the technology enables utilization of electricity instead of fossil fuel derived heat for the separation of succinic acid.

Preferably the electricity used for the separation should be harvested from renewable sources like wind and solar power. These energy sources are inherently fluctuating and an extensive use of these sources will impose a significant stress on the stability of the power grid in the future. This talk demonstrates how a flexible operation of the electrochemical pH-swing extraction process can turn a bio-based succinic acid production into a virtual power plant. We implemented an algorithm in MATLAB that given the price profile of the electricity market calculates the cost-optimal load profile for the electrochemical unit operations. Modulating the power consumption according to the electricity prices provides a 7% saving on the electricity costs. A dynamic model of the electrochemical pH-swing extraction is used to calculate the electric current density, pump-speed and other operational parameters from the dynamic load profile. With the determined operational parameters the load flexible operation of a 100cm² electrolysis cell was tested experimentally. Throughout the experiment a constant degree of protonation close to 100% was measured, indicating that despite the load flexible operation a constant separation efficiency is viable with the operational conditions suggested with the model.

The developed method enables a load flexible and cost optimal operation of the electrochemical pH-swing extraction of succinic acid. In addition to the cost advantage harvested from times of abundant renewable electricity, hydrogen and oxygen are concurrently produced in the electrochemical pH-swing step. The cost advantage together with the generation of valuable co-products constitute an added value that can improve the overall economic competitiveness of the bio-based succinic acid production. Beyond that, load flexible operation and demand side management of large consumers aids in stabilizing the power grid and will likely become more important in the renewable centered electricity supply of the future.

Acknowledgements

The authors gratefully acknowledge the financial support of the Kopernikus project SynErgie (03SFK3L1-2) by the Federal Ministry of Education and Research (BMBF) and the project supervision by the project management organization Projektträger Jülich (PtJ).

Literature Cited

1. Becker J, Lange A, Fabarius J, Wittmann C. Top value platform chemicals: Bio-based production of organic acids. *Current Opinion in Biotechnology*. 2015;36:168–175.
2. Michael McCoy. Succinic acid maker BioAmber is bankrupt. *Chemical & Engineering News*. 2018.
3. AiChE Proceedings, editor. *An Electric Drive for Biorefineries: Electrochemical pH-Shift Liquid-Liquid Extraction of Succinic Acid.*, 2019.
4. Gausmann M, Kocks C, Doeker M, Eggert A, Maßmann T, Jupke A. Recovery of succinic acid by integrated multi-phase electrochemical pH-shift extraction and crystallization. *Separation and Purification Technology*. 2020;240(240):116489.
5. Kiss AA, Lange J-P, Schuur B, Brilman DWF, van der Ham AGJ, Kersten SRA. Separation technology–Making a difference in biorefineries. *Biomass and Bioenergy*. 2016;95:296–309.

632720 Electrokinetic Modeling of Salt Transport and Rejection in Graphene Oxide Nanofiltration Membranes

Abstract:

Graphene oxide (GO)-based membranes have gained tremendous interest in nanofiltration (NF) and reverse osmosis (RO) applications. A major challenge in GO membrane science is to improve their salt rejection in NF and RO processes. Previous researchers have proved that the interlayer spacing between the GO layers as well as the charge density of the GO layers can strongly influence salt rejections from ideal/dilute solutions. However, many practical desalination and dewatering processes operate at much higher concentrations. In this work, our goal is to systematically understand and predict the effect of salts concentration on GO membrane properties by electrokinetic modeling supported by experimental measurements. The correlations between feed concentration, interlayer spacing, and charge density of these GO membranes were determined by single-component salt permeation experiments and ion adsorption models. The modified Donnan steric pore model was then used to model, fit, and predict the salts rejections at different conditions. We then use the transport model investigate in detail the contributions of convection, diffusion, and electro-migration as a function of feed conditions. Finally, the combined effects of initial charge density and interlayer distances on salts rejection are evaluated through the plotting of 2D heat maps. Our study indicates that for neutral pH, under low salt concentrations (~ 0.01 M), modifying the charge density of GO will be more helpful for increasing salt rejections. On the other hand, under high concentrations (above 0.1 M), decreasing the interlayer distance of the GO membranes (i.e., reduction of effective pore size) would be a better strategy. Finally, we show the accurate prediction of binary/multicomponent salt mixture permeation behavior using the model. Overall, this study provides mechanistic insight into salt transport and rejection in GO membranes, and useful guidance for modification of GO membranes in a number of desalination processes.

622103 Rapid and Real-Time Mixed-Plastic Waste Analysis Using Infrared Spectroscopy and Machine Learning

Abstract:

The production of plastics continues to climb but their recycling rate remains quite low. In 2015, nearly 381 million tons of plastic were produced, but only 20% of the world's plastic waste was recycled [1]. Mixed plastic waste (MPW) is putting unprecedented pressure on landfills [2] and it is critical to convert MPW into clean and usable materials to enable a more circular system. A fundamental challenge that hinders this is the difficulty in analyzing the composition of MPW; such information can enable the development of more appropriate treatment strategies.

Traditionally, offline infrared (IR) spectroscopy has been used to identify the chemical components in MPW [3]. Offline IR spectra are easy to identify but are labor- and time-intensive to obtain; as such, this method has limited use in real-time plastic composition assessment. In contrast, online IR spectroscopy allows for rapid detection of plastic composition with a single scan time of 11 microseconds [4]. However, the spectra obtained by this method can be highly noisy (due to interference).

Recently, machine learning methods such as convolutional neural networks have been used to analyze spectral data (in the context of soil characterization) and they have been shown to provide high predictive accuracies and to be resistant to noise [5]. Conventionally, IR spectra are treated as one-dimensional vectors that are analyzed using one-dimensional (1D) CNNs [6]. The 1D CNN works by convolving the spectrum using filters to extract hidden patterns. Although IR spectra already carry a rich amount of information, the hidden correlations between the signal intensities at different wavenumbers are not explicitly encoded.

In this work, we propose to encode the signal correlation information using Gramian angular fields (GAFs) and analyze GAF matrices using 2D CNNs [7]. GAFs have been used to transform time-series vectors into matrices to capture temporal correlations and thus improve classification accuracy. To facilitate fast training of 2D CNN-based algorithms, we use a Piecewise Aggregate Approximation (PAA), which allows us to reduce the dimension of the field matrix [8]. We also used saliency analysis [9] to understand the most important signal intervals and correlations between signals. We show that this approach provides a robust, low-cost, and rapid method for analyzing the composition of MPW and can enable future, high-throughput recycling of plastics.

[1] Ritchie, Hannah, and Max Roser. "Plastic pollution." Our World in Data (2018).
[2] Abdel-Shafy, Hussein I., and Mona SM Mansour. "Solid waste issue: Sources, composition, disposal, recycling, and valorization." Egyptian journal of petroleum 27.4 (2018): 1275-1290.
[3] Zhu, Shichao, et al. "Plastic solid waste identification system based on near infrared spectroscopy in combination with support vector machine." Advanced Industrial and Engineering Polymer Research 2.2 (2019): 77-81.
[4] "S2050 2.0-5.0μm Fiber Spectrometer - NLIR." [Online]. Available: https://www.nlir.com/products/s2050-2-0-5-0μm-fiber-spectrometer/. [Accessed: 15-Mar-2021].
[5] Ng, Wartini, et al. "Convolutional neural network for simultaneous prediction of several soil properties using visible/near-infrared, mid-infrared, and their combined spectra." Geoderma 352 (2019): 251-267.
[6] Chen, Xiaoyi, et al. "1D convolutional neural network for the discrimination of aristolochic acids

and their analogues based on near-infrared spectroscopy." Analytical Methods 11.40 (2019): 5118-5125.
[7] Wang, Zhiguang, and Tim Oates. "Encoding time series as images for visual inspection and classification using tiled convolutional neural networks." Workshops at the twenty-ninth AAAI conference on artificial intelligence. Vol. 1. 2015.
[8] Keogh, Eamonn J., and Michael J. Pazzani. "Scaling up dynamic time warping for datamining applications." Proceedings of the sixth ACM SIGKDD international conference on Knowledge discovery and data mining. 2000.
[9] Sundararajan, Mukund, Ankur Taly, and Qiqi Yan. "Axiomatic attribution for deep networks." International Conference on Machine Learning. PMLR, 2017.

625006 Efficient Biodegradation of Poly(ethylene Terephthalate) with Leaf-Branch Compost Cutinase

Abstract:

Poly(ethylene terephthalate) (PET) is one of the most widely used polyester plastics in the world, but is extremely difficult to be hydrolyzed. Accumulation of the PET in the environment at a staggering rate as discarded packaging and textiles pose a global environmental problem. Cutinases are polyester hydrolases that show a remarkable capability to PET to its monomeric units. This revelation has stimulated research aimed at developing sustainable and green cutinase-catalyzed PET biorecycling methods. The gene encoding a cutinase homolog, LCC, was cloned from a fosmid library of a leaf-branch compost metagenome by functional screening using tributyrin agar plates. LCC is particularly suited toward these ends given its relatively high PET hydrolysis activity and thermostability. With the assistance of the pelB leader sequence, more than 36% of the recombinant LCC produced by *Escherichia coli* BL21(DE3) cells was excreted into the culture medium. To improve the biodegradation efficiency, recycled PET (RPET) was cut into small pieces via grinding to decrease PET particle size and increase surface area. Different particle size of RPET were examined. As a result, the degradation rate of the LCC in the presence of smaller RPET particle size (with 60 mesh screen) was significantly improved, with a productivity of 1.2 milligrams of terephthalic acid (TPA) per milliliter over 48 hours (with enzyme concentration of 0.8 micrograms per milligram of PET). Moreover, the substrate-binding modules from a polyhydroxyalkanoate depolymerase from *Alcaligenes faecalis* (PBM) was fused with the linker (2.3 kDa) from a cellobiohydrolase from *Trichoderma reesei* to the C-terminus of LCC to improve the PET adsorption. Although PBM (6.3 kDa) binding modules have a hydrophobic nature, it was possible to express the proteins in *E. coli*. Our result exhibited that adsorption to PET by the LCC-PBM fusion enzyme was strikingly enhanced as compared with LCC. Though the PET hydrolysis performance of any LCC variants was much higher than that of the bacterial enzyme Is-PETase from *Ideonella sakaiensis* 201-F6, the degradation efficiency of LCC-PBM into TPA was not as high as LCC. This study provides us a great potential to achieve enzymatically depolymerization of PET waste contributing towards the concept of a circular PET economy.

631844 The Use of Life Cycle Assessment to Identify Sustainability Characteristics of Plastic Recycling Systems: A Review

Abstract:

The extensive use of plastics in a "take, make and dispose" order is putting a strain on our environment. Conventional end-of-life treatments of plastics diminishes the potential of waste plastics to extract the maximum value from it through recycle or upcycle. Life cycle assessment (LCA) is an established tool to evaluate technologies of their environmental footprints. A standardized LCA framework enables comparison between technologies and guide the path forward. We have critically reviewed 33 research articles and scientific reports of published LCAs with a view to present an extensive state of the art of the current chemical recycling processes involved in the plastics circular economy system along with their economic and environmental impacts. The overarching goal is to compare environmental implications of technologies for waste to fuel/monomer pathways (such as pyrolysis, and/or gasification) with that of the conventional waste plastic conversion technologies such as incineration and/or landfilling. An array of different waste feedstocks (e.g., both high- and low-density polyethylene, polystyrene, and polyethylene terephthalate) with or without contaminants were analyzed. In the process, the results obtained imply large uncertainties, for example, one study showed mixed plastic incineration was a better option than gasification-pyrolysis due to lower greenhouse gas (GHG) emissions whereas a large number of other studies favored gasification and pyrolysis over incineration due to lower emissions. This work also aims to develop systematic guidance for researchers and practitioners to use LCA to inform decision-making, design, and optimization of recycling pathways and infrastructure, and to support the development of new technologies to advance the plastics circular economy.

Currently, the availability of life cycle inventory (LCI) data remains a major problem. Previous studies usually apply static data inputs (such as using Ecoinvent inventory which is mostly for European data when conducting LCA for US based scenarios), despite the large variations stemming from assumptions of capacity, process design decisions, unit operation performance, etc. To address this gap, lab-scale results coupled with systems design and scale-up must be integrated with LCA. There's also a clear need to provide transparent characterization of uncertainty and the sensitivity of results in modeling environmental impacts of plastic recycling systems, but this is yet to become standard practice in the field, only 15% of the total papers include both sensitivity and uncertainty analysis. Furthermore, use of different impact assessment (IA) methods may present variability in the result due to differences in characterization factors across IA methods under the same impact category. For co-products handling, System expansion is most used in the reviewed papers, with the need to examine how different co-product allocation methods can impact environmental implications. Finally, while addition of TEA strengthens the feasibility of newer technologies in plastic upcycling, very few studies have incorporated both techno-economic analysis (TEA) and LCA together. To elucidate the sustainability implications of technologies via comparative systems analysis, a unified LCA framework is required along with quantifying the technical and economic implications of the system via TEA. Addressing these gaps will help practitioners to make informed and justified methodological decisions on the current upcycling technologies.

620342 The Nano-F-Pores Process for the Production of Nanocarbons Form Waste Plastic

Abstract:

This work presents the scale-up endeavour of the novel Nano-F-PoRes process aimed at producing carbon nanofilaments (CNF) and synthesis gas from waste plastic. The pilot scale process (kg-lab) has a CNF production capacity of 1 kg/h, which is roughly 250 times higher than the current g-lab one. Although such a capacity is still lower than what can be qualified as a commercial unit, it could be easily used as a pilot unit towards commercialization.

The methodology involves:

- A continuous bubbling fluidized bed (BFB) autothermal pyrolyzer (ATP), characterized by two in series regimes (oxidative and reductive), is designed, built and presently tested to optimize its operation. The BFB can be either an inert particulate material (i.e. olivine or alumina or silica sand) or a component contributing as thermo-catalytic cracking and reforming material. Preliminary tests at g-lab scale are available and will be presented.
- The product of the ATP step is fed into a hybrid catalytic reactor vessel combining a bubbling fluid and a mobile catalytic bed (HMFBR); this vessel was patented as a mobile bed filter and it is now tested as a reactor vessel for the production of CNF. At lab scale the CNF are produced in a semi-batch reactor composed of a catalytic fixed bed under the form of metal sheets appropriately positioned in the reactor space (vertical star–valve configuration). Such a configuration allows for a uniform distribution of the reactants and maximizes the CNF production radially and axially. Reactor must be stopped when CNF fill the metal sheets interspace up to a maximum level defined by pressure loss increase across the bed. The reaction is endothermic and heat must be provided with uniformly. At small scale this heat transfer is simple and it is done electrically. At larger scale heat transfer and semi-batch operation become major challenges. The catalytic reaction producing CNF has been tested at g-lab scale with various pyrolysis products surrogate molecules and the results will be also presented.
- The so-produced CNF are entrained along with the gaseous products of the HMFBR and they are recovered continuously using a particles retention train based on the utilization of high T metallic filters.

The pilot is installed in a recently inaugurated process scale-up building at the Université de Sherbrooke. 3D drawings and pictures of the equipment as well as the so far evolution of the project and the results will be presented.

Since the process is presently at the stage of submitting a patent, this abstract cannot include more data. The latter will be though available to be presented during the conference.

629417 Design and Simulation of Thermo-Chemical Plastic Waste Recycling to Produce Di-Methyl Ether (DME)

Abstract:

The extent of plastics use in everyday life is evident by the fact that the annual global plastic production has exceeded 350 million tons. Most of the plastic materials ends up in landfills after their use and approximately 3 % of global plastic production has littered the oceans causing negative impacts on the environment and marine life. The projected continuous increase in the production of plastics due to their versatile applications also represents an economic opportunity by recycling them while addressing the adverse environmental concerns. Currently employed mechanical recycling techniques have limitations because of poorer product quality after each cycle. Hence, chemical recycling of plastic waste has been an active area of research recently.

In this study, we investigated the thermochemical plastic recycling processes namely, gasification. This study rigorously modeled and simulated the thermochemical processes that can recycle the plastic wastes efficiently to useful chemical such as di-methyl ether (DME). Currently, the cost of recycled plastic is not competitive compared to the fossil-based virgin plastics, which leads to the lack of industrial willingness for recycling. This study explored the technical and economic factors that can improve the economics of the recycled plastics. The design of reactors followed by the separation train was conducted to decrease exergy losses and improve the overall efficiencies. The results show that the gasification can provide a competitive solution with flexibility in terms of waste feed composition to produce DME at a large scale.

622178 Estimating Plastic Flows in the U.S.: Processes for Recycling PET

Abstract:

Post-consumer use plastic can be employed as a feedstock in chemical processes, but first one needs an understanding of the amount and purity of these recycle streams. The same information supports the U.S. Environmental Protection Agency's report on Advancing Sustainable Materials Management: Facts and Figures. This report includes information on municipal solid waste (MSW) generation, recycling, composting, combustion with energy recovery, and landfilling. Plastics have been identified as key components in post-consumer material flows and worthy of tracking. In this effort throughputs for material recovery facilities (MRFs) around the country are identified with the intent of modeling input-output flows. Polyethylene terephthalate (PET) is a plastic of particular interest because it is the most widely recycled. PET is used in plastic bottles, thermoforms for food packaging, carpeting, and textiles, but only PET bottles are regularly recycled. Post-consumer PET flows through MRFs and specific recycling processes. These processes use resources and produce environmental releases, which along with the recycled flow amounts and purities describe the PET recycling system, finally leading through chemical engineered conversion into recycled PET products. This system has the potential to provide an example of a circular economy, where materials are kept in use and only downcycled as appropriate.

The views expressed in this presentation are those of the authors and do not necessarily represent the views or policies of the U.S. Environmental Protection Agency.

631077 Technoeconomic Analysis of Torrefied Pellets Produced from Municipal Solid Wastes

Abstract:

The U.S. Department of Energy promotes the production of advanced biofuels through support for fundamental and applied research pertaining to feedstock logistics and preprocessing. The disposal of municipal solid wastes (MSW) presents a growing challenge for municipalities and similar entities and transforming these waste streams to higher-value products presents both an economic opportunity and a sustainable solution by diverting materials away from landfills. Furthermore, identifying and developing advanced preprocessing systems is also imperative for MSW resources to be utilized at scale. This research evaluates the production of torrefied pellets from waste plastic, fiber and biomass residues, which can be utilized as fuel to produce electricity, from a technoeconomic standpoint. The major benefit of the studied preprocessing train is the elimination of bound chlorine, increased energy density during densification to a storage fuel pellet and increase in product uniformity (physical and chemical) at both the macro- and sub-millimeter scale. The system boundary for this analysis encompasses processes and equipment beginning from delivery of waste materials, size reduction using a three-stage shredding process, torrefaction in a reactor-extruder, and cooling of processes materials to produce 100,000 dry tons of torrefied materials annually. The technoeconomic analysis estimated the costs and energy consumption for the MSW torrefaction and pelleting operations. The analysis indicated that the baseline cost of processing sorted MSW material into uniform pellets is about $50.15/dry ton (2020$), which includes preprocessing costs at the equipment-level and other plant-level costs comprising of installation, labor, and maintenance. The most important preprocessing costs are attributable to the crammer and torrefier ($9.14/dry ton) and the three-stage shredder ($12.66/dry ton). At the plant-level annual labor costs ($15.07/dry ton) and installation costs ($8.62/dry ton) are the largest contributors to total costs. The system is designed to be self-sustaining whereby the gas stream from the torrefaction process is utilized in the heat management system with minimal usage of natural gas in the start-up phase. This study contributes to the growing body of research in the areas of waste utilization and management, energy recovery, and sustainability by evaluating novel production pathways for biopower application and to improve the economic potential of biopower production and use.

626502 Global Prediction of Mass Distribution of Plastic Waste in Environment

Abstract:

In recent years, plastic pollution in environment has become a serious problem all over the world. To assess environmental impacts of plastic waste accumulation in environment, crucial steps are to quantify the amounts of plastic lost to the environment from different sources and fluxes between environmental compartments and accumulation areas for plastics. This study represents the first quantitative global prediction of the mass distribution of plastics across environmental compartments. Probabilistic material flow analysis was used to obtain accumulated mass of plastic in various environmental compartments, based on well-known sources of plastic release to the environment and transport pathways between environmental compartments. The analysis of plastic released to the environment based on a quantitative model considering all relevant sources and transport pathways across environmental compartments are the difference our work from previous work. The model estimates the mass distribution of plastic in aquatic and soil environments including fragmentation of larger plastic to smaller pieces (i.e. macroplastics to microplastics and microplastics to nanoplastics). We estimated that 9.6±4.7 Mt of plastic (i.e. plastic excluding rubber and road marking) were lost to the environment of which 4.5±1.9, 3.3±1 and 1.4±1 Mt /year accumulated in ocean, soil and freshwater system, respectively, which is first estimation of distribution of plastics in the environment. Furthermore, 0.4±0.8 Mt of plastic was estimated to be degraded in environment. In addition to this, 0.6±0.2 Mt and 1±0.6 Mt were accumulated in agricultural soil and ocean floor respectively. Among freshwater compartments, freshwater sediments (floor and shorelines) were estimated to accumulate 0.9±0.6Mt of plastic. The ocean shorelines (30% of plastics) and urban soils (27 % of plastics) were predicted as hotspots for plastic accumulation. This study can be considered as a baseline data for quantifying plastic transport and accumulation areas, which can be linked to future ecotoxicity studies and risk assessments of plastic waste.

630032 Understanding the Phase Behavior of Conjugated Polymer Composites By Using AFM-IR

Abstract:

In this talk, I will discuss the use of AFM-IR to quantify the phase behavior of various polymer composites. I will first discuss the background to the tool, then transit to the new conjugated polymers and elastomers composites enable new materials property in stretchable electronics.

625192 Fabrication of Polystyrene/Graphene Quantum Dots Nanocomposites for Enhanced Flame Retardant Properties *Via* Pickering Emulsion Polymerization

Abstract:

Significant improvements in thermal and flame retardant properties of polymeric materials holds tremendous promise for fire safety materials. Carbon-based nanoparticles have low cytotoxicity, chemical inertness, cost-effectiveness, and biocompatibility. In the present work, an alkyl modified graphene quantum dots (C-GQDs) were deposited on a polystyrene (PS) sphere (PS/C-GQDs) for reducing its fire hazards. The PS/C-GQDs nanocomposites were fabricated *via* Pickering emulsion polymerization directly in the presence of the amphiphilic C-GQDs was as sole stabilizer during the polymerization process. Morphological characterization by SEM suggested the successful preparation of C-GQDs stabilized PS colloidal particles, and C-GQDs layer was wrapped on the surface of PS microspheres. It could provide a potential pathway to introduce flame retardant to polymeric materials directly and efficiently. Enhanced thermal stability and flame retardancy were obtained for PS/C-GQDs which synthesized with 2% C-GQDs in aqueous phase. There is a decrease *ca.* 35% in peak heat release rate compared to neat PS, respectively. The enhanced fire-retardant properties should be ascribed to the barrier of assembled C-GQDs on sphere surface. And the incorporation of C-GQDs could remarkably improve the stability of char layer. And the char layer served as an efficient insulating barrier to impeded the permeation of heat and the escape of pyrolysis volatile products. Thus, this work paves a potential pathway to design efficient assembled fire-retardant systems for fire safety.

624995 Self-Assembly and Fibrillization of Waste Protein into High-Performance Films

Abstract:

Petroleum resources have greatly improved longevity and quality of life across the globe through innovative new polymers, but they have also contributed to the escalating crisis of global climate change and plastic waste in the environment. Therefore, there is an urgent need to develop alternatives that can enhance sustainability. While a great deal of work has focused on sugar-based materials, proteins offer a promising feedstock to mimic and replace polymers such as nylon and polyurethane. Polyurethanes are a particularly attractive target for replacement because of their additional chemical hazards and toxicity in the manufacturing process. However, proteins are usually either too brittle (neat materials) or too weak (plasticized materials) to serve as suitable substitutes. Our group has previously demonstrated that this limitation can be overcome by designing protein copolymers in which proteins act as 'hard blocks' and low-T_g acrylates as 'soft blocks'. In order to achieve broader applications, however, there are other scientific challenges, for example, the reported range of mechanical properties does not yet match that of high-strength synthetic polyurethanes.

Since protein domains ('hard blocks') in current materials are largely amorphous, an increase in protein crystallinity (i.e., formation of ordered structures) is expected to improve mechanical performance. Amyloid fibrils prepared from protein monomers stand out because they are the strongest known proteinaceous materials, owing to their highly ordered and dense cross-b sheet structures. A large number of different proteins, including those derived from food and agricultural waste such as whey protein, have been shown to be readily converted into amyloid nanofibrils *in vitro* under designed conditions. Here, 'hard block' crystallization was improved by carefully controlling protein fibrillization, leading to the mechanical improvement of the resultant materials. We chose whey protein isolate-hydroxypropyl acrylate copolymers as an example and found that protein fibrillization can significantly improve the mechanical properties of the ensued materials. In particular, compared to the protein monomer system, protein fibrillization led to an 8, 3, and 2-fold increase in Young's modulus, tensile strength, and toughness of the final elastomer, respectively, with minimal reduction in strain at break. This result provides a scalable and sustainable strategy for improving the performance of protein materials and replacing polyurethanes-like materials.

632482 Burn-Dry: Aerogel Fabrication By Polymer-Assisted Rapid Thermal Annealing

Abstract:

Aerogels, comprised of a microporous solid in which the dispersed phase is a gas, are desired for a broad range of emerging applications, such as super insulation and energy-saving. These rising needs highlight the importance of developing a simpler, scalable strategy for aerogel fabrication that by-passes uneconomical steps, including long thermal treatments, etching, and freeze-drying. In this work, we leverage rapid thermal annealing (RTA) to fabricate reduced graphene oxide (rGO) aerogel as a new "burn-dry" method that is more efficient and potentially applicable to various materials. The RTA is a classic method in the semiconductor industry to dope silicon by heating up wafers high temperatures (over 1000 °C) at a very fast rate (up to ~150 °C/s). Unlike extensive research on semiconductors, applications of RTA to polymers are less explored, perhaps because the thermophysical properties of polymers under such high heating rates are not well understood. Polymer-assisted RTA can be considered as a burn-dry, as opposed to freeze-drying, which freezes structure by quenching and removing solvent with tedious steps. In contrast, polymer-assisted RTA quickly burns polymer away, thereby preventing molecular rearrangement to freeze the structure. We hypothesize that the very fast heating rates of RTA can kinetically arrest the co-continuous structures to template a model hard material, rGO. This simple, fast, energy-efficient technique has already been employed in the semiconductor and solar cell industry, making it versatile, highly scalable, and economically viable. This presentation will discuss the relationship between heating rate and phase separation kinetics of a model polymer blend with rGO and resulting morphology.

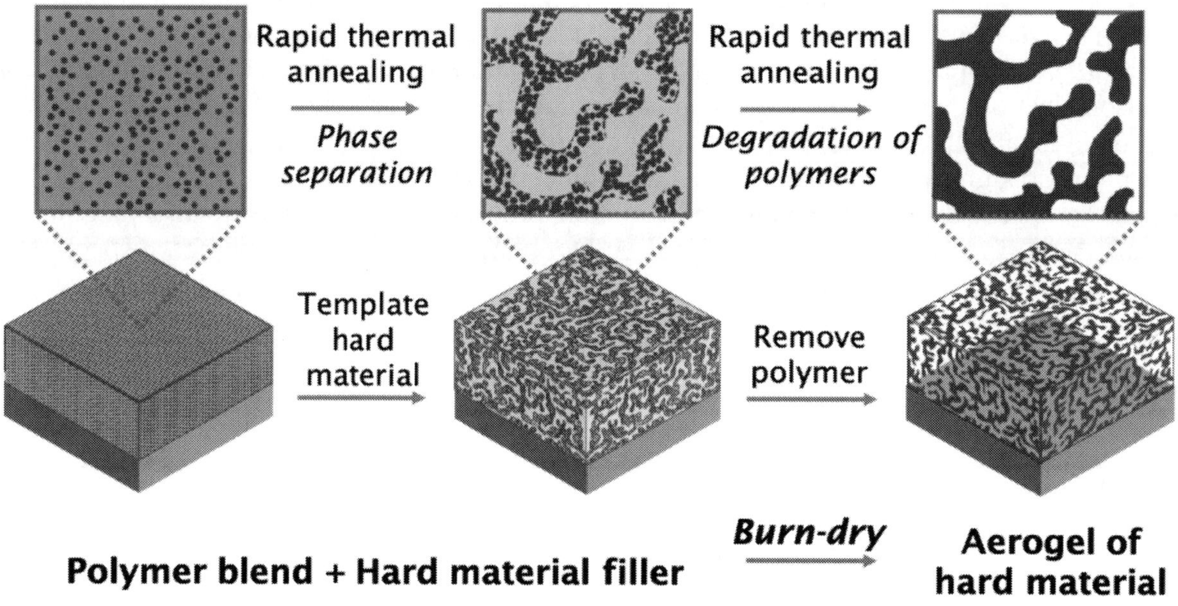

625755 Gill-Structured Composites for Sustainable Energy Collection and Self-Powering Wearable Devices

Abstract:

Thermoelectricity fills the niche market of long-term energy supply and waste heat recovery by directly converting thermal energy into electricity. To harvest energy from curved surfaces, such as human skin, mechanical flexibility and novel structure are needed in a thermoelectric generator (TEG) to build a high-temperature gradient. Conjugated polymers have the potential to act as thermoelectric materials due to their electrical conductivity, accessibility, non-toxicity, and low thermal conductivity. Herein, we report available multiwalled carbon nanotubes (MWNTs) and polyaniline (PANi) composites as inexpensive thermoelectric materials. *In-situ* polymerization took advantage of the π- π stacking between PANi and MWNTs, and optimized their interface to reach a high power-factor and improved mechanical robustness. Additive manufacturing was conducted to obtain a gill-mimicking flexible substrate which formed the PANi/MWNTs hybrids in an out-of-plane manner. The simulated temperature distribution showed the better ability of the gill-structured generator to yield high-temperature differences over a flat design. Extraordinary bendability and durability were shown in 1000 bending tests. Our demonstration presented stable voltage generated by the TEG on hot water containers, hot car roofs, and human skins. Moreover, the TEG can power a sensing system that reveals biosignals, such as joint movement and respiration, with a strain sensor. This project sheds light on health monitoring, wearable electronics, Internet-of-things, and human-computer interaction.

621817 Effects of Zeolitic Imidazolate Frameworks (ZIFs) on Fire Behaviors of Intumescent Flame-Retarded Polypropylene (PP) Composites

Abstract:

As an efficient halogen-free compound, the intumescent flame retardants (IFRs) have gradually been commercialized and widely used in industries for a fire-safe polymeric composite. However, there are still some challenges for IFRs, such as poor stability and low dispersion efficiency in the polymer matrix. A combination of conventional IFRs with metal organic frameworks (MOFs) is a promising strategy to develop well-dispersed and highly efficient flame-retarded polymer nanocomposites. In this work, ZIF8-IFR-PP polymeric nanocomposites were synthesized, characterized, and followed by various fire testing. It was found that when adding 2 wt% ZIF-8, the peak heat release rate was decreased to 172.90 kW/m^2 compared with 298.30 kW/m^2 for IFR-PP and 904.98 kW/m^2 for neat PP. ZIF-8 nanocomposite also shows an excellent smoke suppression. A reduction of total smoke release was observed from 2704.2 m^2/m^2 to 1886.2 m^2/m^2 and peak smoke production rate decreased from 0.10 m^2/s to 0.045 m^2/s. Another released toxic component CO production rate were also suppressed during the combustion period. The reasons for these improvements are ascribed to a synergistic effect between IFR and ZIF8 on promoting cross-linking reaction in the condense phase for a carbonaceous physical barrier. Different metal irons (ZIF67) and bimetal ZIFs (ZnCo-ZIF) were also synthesized to further investigate how metal iron compositions affect the fire safety performance of ZIF-IFR-PP nanocomposites. Interestingly, it was found that ZIF67 has promoted more stable crosslinking components than ZIF8, which leads to a more compact and continuous char structures. CO production rate during the combustion was further decreased in ZIF67 and ZnCo-ZIF nanocomposites compared with ZIF8 nanocomposite due to the catalytic oxidation process from the cobalt. Overall, with a small loading (2%) ZIFs in IFR-PP, it can remarkably reduce the flammability and toxicity during the polymer combustion, including heat release rate, smoke production, and CO production rate. Therefore, the fabrication of MOFs together with conventional flame retardants provide new opportunities on developing novel fire-safe polymeric composites.

630595 Modeling Evaluation for Pyrolysis of Biomass and Plastic Waste

Abstract:

Availability of biomass and plastics is increasing day by day and along with it is increasing the availability of waste plastic and biomass. This waste is required to be treated judicially to make sure, the waste will not be hampering the environment and to utilize it in an efficient way. With the advancement in technology, and the requirement of the fuels for various purposes, the conversion of the biomass and plastic to fuels is gaining importance in research and applications. Gasification proves to be an enriched approach for converting the waste to electrical sources. On the similar front, the research is also gaining a lot of importance in the said field.

The present work focuses on the evaluation of the different processes applicable in the waste treatment be it the plastic or the biomasses. The modeling approach is considered using the Aspen Plus platform which simulates the sub-processes of the gasification for the implementation. The pyrolysis and gasification process are modeled utilizing the experimental data and regression. The review shows that the higher power generation can be obtained on comparing the biomass content in the combination of biomass and plastic wastes. It has also evaluated the pyrolysis process for converting of the plastics and other biomass waste to the useful energy and fuels ultimately leading to the utmost profitable and environmental benefits.

AUTHOR INDEX